MW01378879

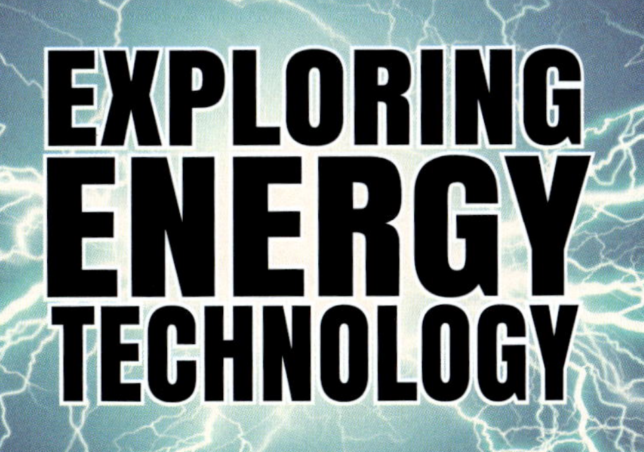

FOSSIL FUELS

EDITED BY ELIZABETH LACHNER

~~The John P. Holt Brentwood Library~~
~~8109 Concord Road~~
~~Brentwood, TN 37027~~
~~Phone: 615-371-0090~~

Published in 2019 by Britannica Educational Publishing (a trademark of Encyclopædia Britannica, Inc.) in association with The Rosen Publishing Group, Inc.
29 East 21st Street, New York, NY 10010.

Copyright © 2019 by Encyclopædia Britannica, Inc. Britannica, Encyclopædia Britannica, and the Thistle logo are registered trademarks of Encyclopædia Britannica, Inc. All rights reserved.

Rosen Publishing materials copyright © 2019 The Rosen Publishing Group, Inc. All rights reserved.

Distributed exclusively by Rosen Publishing.
To see additional Britannica Educational Publishing titles, go to rosenpublishing.com.

First Edition

Britannica Educational Publishing
J.E. Luebering: Executive Director, Core Editorial
Andrea R. Field: Managing Editor, Compton's by Britannica

Rosen Publishing
Amelie von Zumbusch: Editor
Brian Garvey: Series Designer/Book Layout
Cindy Reiman: Photography Manager
Nicole DiMella: Photo Researcher

Library of Congress Cataloging-in-Publication Data

Names: Lachner, Elizabeth, editor.
Title: Fossil fuels / edited by Elizabeth Lachner.
Description: First edition. | New York : Britannica Educational Publishing, in Association with Rosen Educational Services, 2019. | Series: Exploring energy technology | Audience: Grades 5–8. | Includes bibliographical references and index.
Identifiers: LCCN 2018006605| ISBN 9781508106135 (library bound) |
ISBN 9781508106128 (pbk.)
Subjects: LCSH: Fossil fuels—Juvenile literature.
Classification: LCC TP318.3 .F67 2019 | DDC 662.6—dc23
LC record available at https://lccn.loc.gov/2018006605

Manufactured in the United States of America

Photo credits: Cover spooh/E+/Getty Images; cover, back cover and interior pages (background) © iStockphoto.com/tolokonov; pp. 4-5 Rudmer Zwerver/Shutterstock.com; p. 7 ekipaj/Fotolia; pp. 9, 13, 21, 23, 30 © Encyclopædia Britannica, Inc.; p. 12 © AP Images; pp. 14-15 Nick Suydam/Alamy Stock Photo; p. 19 Anan Kaewkhammul/Shutterstock.com; p. 20 (background) sakkmesterke/Shutterstock.com; p. 24 chaiwattudsri/Fotolia; p. 26 Joe Sohm/Visions of America/Universal Images Group/Getty Images; p. 28 ulga/Fotolia; p. 32 William Campbell/Corbis News/Getty Images; p. 33 Ken Cedeno/Corbis News/Getty Images; p. 36 Karen Kasmauski/Science Faction/Getty Images; p. 38 Exxon Valdez Oil Spill Trustee Council/NOAA; p. 40 Sergiy Serdyuk/Fotolia; p. 41 Bastian Kienitz/Shutterstock.com.

CONTENTS

INTRODUCTION

A fossil fuel is a natural substance formed from the buried remains of ancient organisms that can be used as a source of energy. Fossil fuels formed over millions of years as heat and pressure from layers of sediment changed the decayed organic remains into materials such as coal and petroleum. The energy in fossil fuels is the energy from sunlight stored in the tissues of the buried organisms as a result of photosynthesis.

Fossil fuels may be solids, liquids, or gases. All fossil fuels are hydrocarbons, a class of chemicals composed only of carbon and hydrogen atoms. Coal, petroleum, and natural gas are the most commonly known fossil fuels.

All fossil fuels can be burned in air or with oxygen derived from air to provide heat. This

This power plant is burning coal to generate electricity. Like natural gas and oil, coal is a fossil fuel.

heat may be employed directly, as in the case of home furnaces, or used to produce steam to drive generators that can supply electricity. In still other cases—for example, gas turbines used in jet aircraft—the heat yielded by burning a fossil fuel serves to increase both the pressure and the temperature of the combustion products to furnish motive power.

Fossil fuels are not distributed evenly across the Earth. The United States, Russia, and China have the largest coal deposits in the world. Sizable deposits also are located in Australia, India, and South Africa. More than half of the world's known oil and natural gas reserves are located in the Middle East. This means that the Middle East contains more oil than the rest of the world combined. Major reserves are also found in Canada and the United States, Latin America, Africa, and parts of Russia, Transcaucasia, and Central Asia. Each of these regions contains less than 15 percent of the world's proven reserves.

Fossil fuel usage has steadily increased since the Industrial Revolution. At the start of the twenty-first century, fossil fuels comprised nearly 90 percent of the world's energy supplies. However, fossil fuels are nonrenewable resources. Because it takes millions of years for fossil fuels to form, they cannot be replaced when they are used.

Burning petroleum and coal releases harmful gases such as carbon monoxide, nitrogen oxide, and sulfur dioxide. These gases pollute the air and react

Before they can be used, fossil fuels must be extracted from the ground. This beam-pumping unit on top of an oil well is used to lift oil up to the surface.

with moisture in the atmosphere to create acid rain. Scientific evidence shows that burning fossil fuels increases atmospheric temperatures. This warming of Earth's atmosphere, called the greenhouse effect, contributes to climate change, a serious environmental concern.

CHAPTER ONE

COAL

One of the most important natural fuels, coal was formed from plant life buried in the Earth millions of years ago. A hydrocarbon, coal is classified in ranks, or types, according to the amount of heat it produces. This depends upon the amount of fixed carbon it contains. The ranks, in increasing order, are lignite, or brown coal; subbituminous coal, or very soft coal; bituminous coal, or soft coal; and anthracite, or hard coal. Bituminous coal is the most abundant type.

Coal is most commonly used to produce electricity in power plants. It also is an important fuel for heating and powering industrial and manufacturing facilities, and for making steel. The many chemicals derived from coal are used in industrial processes and in the manufacture of nylon, paints, plastics, synthetic rubber, aspirin, and thousands of other useful products.

WHERE COAL IS FOUND

China is estimated to have the largest recoverable coal reserves in the world—about 45 percent of the world's total. The United States has about 23 percent of the total. The United States recoverable reserves are estimated at 265 billion tons (240 billion metric tons)—enough to last about 250 years at present consumption rates.

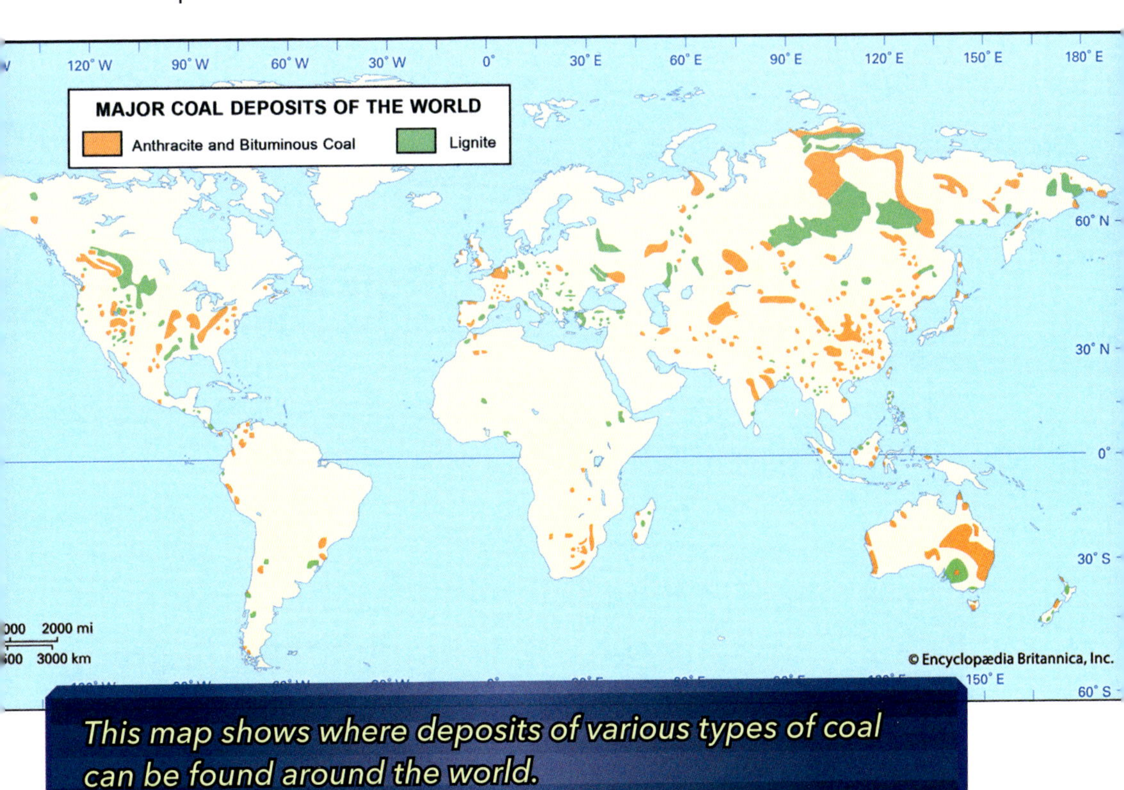

This map shows where deposits of various types of coal can be found around the world.

There are about 2,500 coal mines in the United States. The productivity of these mines, measured in tons per miner per hour, has nearly tripled over the past 30 years, because of increased use of modern mining machinery. The total value of the more than 900 million metric tons of coal produced annually in the United States averages over $20 billion. The chief coal-producing states are Wyoming, Kentucky, West Virginia, Pennsylvania, Illinois, and Texas. Anthracite is mined only in Pennsylvania, but other states have small deposits.

Electric utilities are the major consumers of coal in the United States. The second largest US market is industrial and retail coal use, followed by steel production. US coal exports exceed 100 million tons annually, with a value of about $4 billion.

HOW COAL FORMED

The world's coal beds have resulted mainly from the abundance of trees, ferns, and other plants that existed in the tropiclike forests of the Mississippian and Pennsylvanian geological periods. These periods—called Lower and Upper Carboniferous in Europe and commonly known as the Coal Age—lasted for approximately 60 million years, from about 358 million to 298 million years ago. The major high-quality coal deposits are found primarily in the strata, or layers, of the Pennsylvanian period. After plant life died, it fell into swamp water where it partially decomposed into a slimy, colloidal mass and formed peat. As the seas advanced and receded in cycles over the Earth, they deposited heavy layers of sandstone, shale, and other rocks on top of the peat.

The increased pressures and heat that resulted from the overlying strata caused the buried peat to dry and harden into lignite, a low-grade coal. Under the

OTHER SOLID FOSSIL FUELS

Peat and coke are solid fossil fuels that are commonly used today. Peat is used as a heating fuel in areas where other fuels are not available. It is not an efficient fuel because it burns slowly, producing much smoke and little heat. Coke is a residue that remains after gases and tar are extracted from some types of coal. It is useful in industry because it produces intense heat without smoke. Coke is widely used in blast furnaces to make iron and in other metallurgical processes.

pressure of still more layers, the lignite became subbituminous and bituminous coal. Anthracite, or hard coal, was formed from the bituminous when great pressures developed in folded rock strata during the creation of mountain ranges. This occurred only in certain areas, such as the Appalachian region of Pennsylvania.

MINING COAL

How coal is mined depends mainly on how deep the coal bed lies from the surface and on local geologic conditions. If the coal is within 200 feet (60 meters) or less of the surface, the mine will be a surface, or open-pit (also called open-cut), mine. Surface mines

This mine is in Yukon, West Virginia. In the United States, the average depth of coal mines is 250 to 300 feet (75 to 90 m), but some mines are as much as 2,000 feet (610 m) deep.

generally yield a greater average tonnage of coal per man-hour. Once a coal deposit has been selected for mining, several years of planning and development are necessary before actual extraction can begin. Permits must also be obtained from local and federal agencies.

If the coal is in hilly terrain or is too deep for surface mining to be feasible, an underground mine will be opened. A shaft, called the mine entry, must be dug to reach the coal seam. The entry has three purposes—to take workers and equipment into the mine,

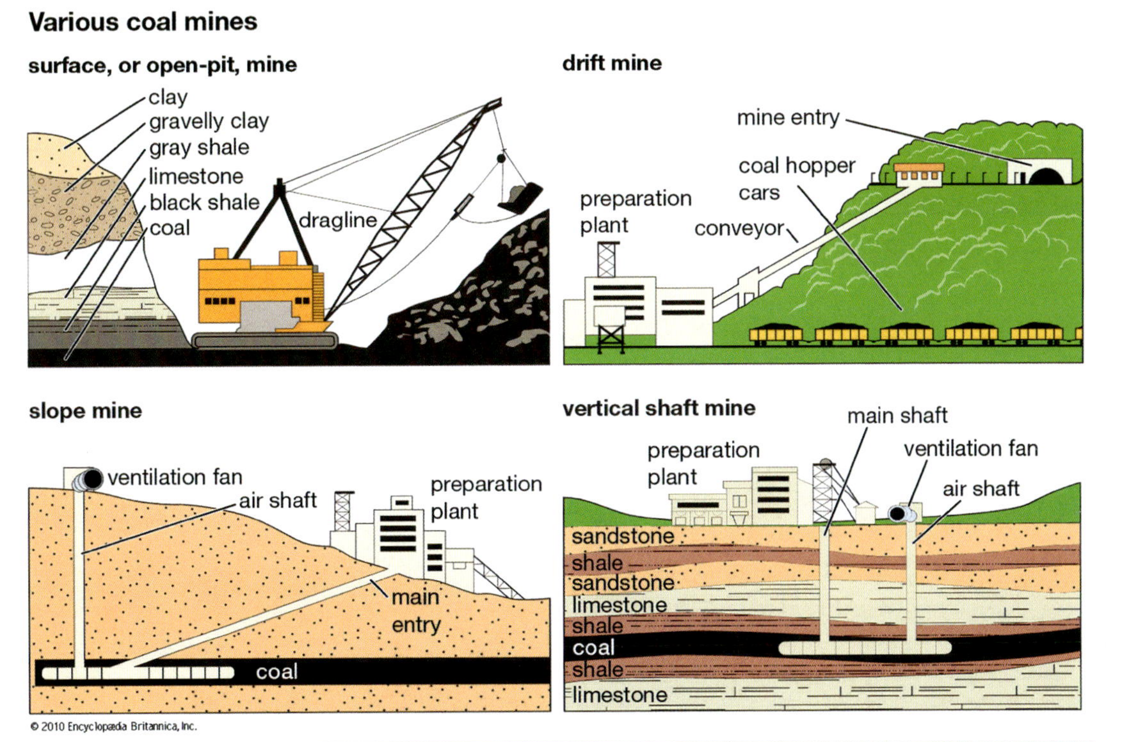

© 2010 Encyclopædia Britannica, Inc.

Where feasible, surface mining is about three times more productive than underground mining. It's easier to use large excavating equipment closer to the surface.

to transport the coal out, and to ventilate the mine.

When a tunnel is driven horizontally into the side of a hill to reach the coal, the mine is called a drift mine. A slope mine has an entry that slopes upward or downward to the coal seam; it must also have adjoining vertical shafts for air ventilation and for emergency exit use. Shaft mines are opened in flat terrain by driving vertical shafts to the coal.

THE PREPARATION AND DISTRIBUTION OF COAL

Raw coal contains rock and other impurities, such as iron sulfide, which are greatly reduced in various cleaning operations. The coal is also crushed, sized, and blended. A preparation plant for bituminous coal is called a tipple. Anthracite goes to a building called a breaker.

In the United States, railroads haul about 60 percent of the coal transported to market. To handle the millions of tons of bulk shipments, railroads use unit trains of

This ship is being loaded with coal in Chicago, Illinois. The ship will travel up the Calumet River to the nearby Lake Michigan.

coal hopper cars that can deliver more than 10,000 tons (9,072 metric tons) of coal at a time to large consumers.

Coal is shipped by barges over inland waterways and along the Atlantic coast, which accounts for more than 25 percent of total tonnage shipped.

Ships called colliers carry coal on the Great Lakes, along the coasts, and for export overseas. A growing percentage of coal output moved over shorter distances, usually 100 miles (160 kilometers) or less, is handled by trucks.

Some electric utility companies burn coal in power-generating plants built near the mines. The energy from these plants reaches the market in the form of electric power over high-voltage and extra-high-voltage transmission lines. An experimental 750,000-volt system using this method—called "coal by wire"—has been developed.

Chapter Two

PETROLEUM

Petroleum is crude oil, a naturally occurring liquid that can be refined to make gasoline, diesel fuel, jet fuel, home heating oil, lubricating oil, wax, asphalt, and many other valuable products. Crude oil usually occurs in subsurface deposits, but in some areas it leaks onto the surface in oil seeps. As a technical term, "petroleum" also includes natural gas, a naturally occurring gas with a similar chemistry to crude oil that also occurs in subsurface deposits. The subsurface deposit of crude oil is called the oil reservoir, and the surface of the ground above it is called the field. The field is given a geographical name, such as the Prudhoe Bay Oil Field.

The main liquid fossil fuels used today are refined products of crude oil. These include gasoline, fuel oils such as diesel and jet fuel, and furnace oils for home heating. Kerosene, once widely used to provide light, is still used in many places for cooking and space heating. It also is the main fuel for modern jet engines.

ORIGINS, DEPOSITS, AND RESERVES

Most of the world's large natural gas and crude oil deposits formed during the Carboniferous Period. Oil and gas formed through a similar process, often in the same swampy location, from the buried remains of tiny aquatic organisms such as algae, diatoms, and zooplankton. As these organisms died and sank to the muddy bottom, decomposition and other changes gradually altered their buried remains into a substance called kerogen. Over millions of years, increasing heat and pressure from more sediment layers transformed the kerogen into petroleum. Depth and temperature determined whether the petroleum was liquid or gaseous, with natural gas forming at deeper and hotter locations.

For a commercial deposit of crude oil or natural gas to occur, there must be a source rock to generate gas and oil, a reservoir rock to hold the gas and oil, and a trap on the reservoir rock to concentrate the

Crude oil is another name for liquid petroleum. The word "petroleum" comes from Latin, in which petra *means rock and* oleum *means oil.*

gas and oil. A source rock is sedimentary rock with organic matter that produces crude oil and natural gas. After oil and gas form in the source rock, they usually flow upward toward the surface. This happens because the sedimentary rocks also contain water, and the oil and gas are lighter in density than water. As they rise up, the oil and gas can flow into a type of sedimentary rock called a reservoir rock. A reservoir rock has billions of tiny, interconnected

UNCONVENTIONAL CRUDE OIL SOURCES

Tar sands (also called oil sands) are a mixture of sand, very heavy oil (tar), and water. Large deposits exist in Canada (the Athabaska Tar Sands in Alberta) and Venezuela (the Orinoco Oil Sands). These deposits are as large or larger than the world's remaining conventional oil reserves. Because tar sands are very viscous, the challenge is to extract the oil from them in a cost-effective way.

Oil shales are another source of unconventional crude oil. The organic matter in oil shales occurs in an intermediate state called kerogen. When heated to 662 degrees Fahrenheit (350 degrees Celsius), kerogen forms light, high-quality crude oil called shale oil. The largest oil shale deposits are located in the United States, Russia, and Brazil. There is more crude oil in oil shales than in all conventional oil and gas reservoirs. However, production has been limited because it is expensive to mine and heat the oil shale and dispose of the waste shale and water without having a major environmental impact.

spaces called pores. When a well is drilled into the rock, water, gas, and oil can flow from pore to pore through the rock and into the well.

Crude oil and natural gas are concentrated in subsurface traps, which are high areas on reservoir rocks. Because the pores of the reservoir rock are filled with water, the lighter crude oil and natural gas flow to the highest area. A cap rock or seal—a sedimentary rock that does not allow fluid to flow through it—must overlay the reservoir rock in the trap.

Reserves are the amount of crude oil that is expected by engineering calculations to be produced in the future from both conventional and unconventional sources. In the early twenty-first century

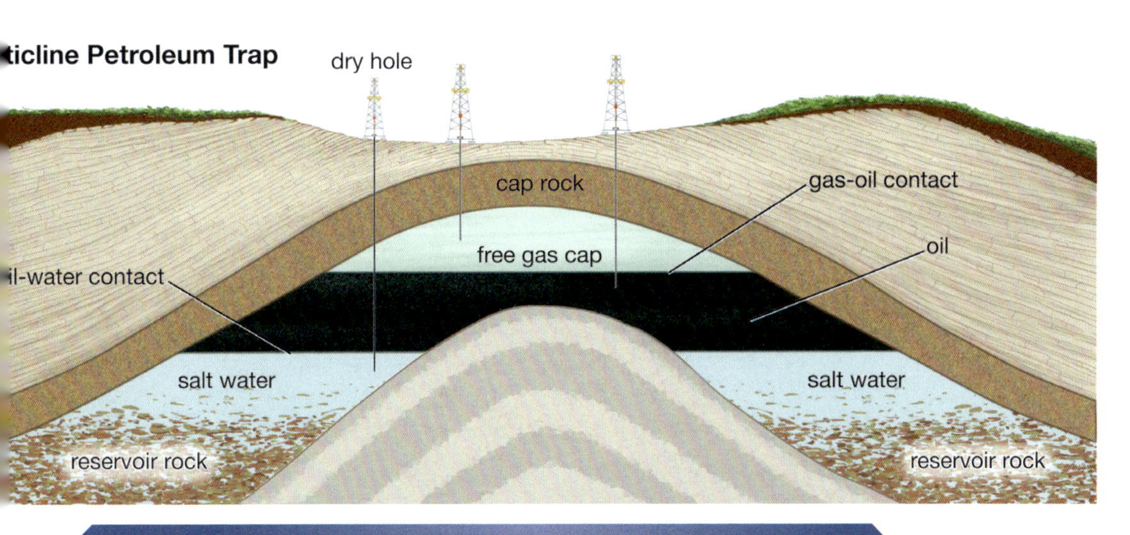

The most common type of structural trap is formed by an anticline, a structure with a concave roof caused by the deformation of the reservoir rock and the impermeable cap rock.

there were about 1.3 trillion barrels (200 billion cubic meters) of remaining oil reserves worldwide. The annual world consumption of oil is about 30 billion barrels (4.8 billion cubic meters) each year. More than 1 trillion barrels (150 billion cubic meters) of oil have already been consumed. More than half of the world's oil reserves—56 percent—are in the Middle East.

PETROLEUM EXPLORATION

Geologists study rocks to find crude oil. They map how sedimentary rocks crop out on the surface, which can indicate the presence of subsurface traps such as anticlines and domes. To identify subsurface source rocks and potential reservoir rocks, geologists use data from wells that have already been drilled. They match rock layers between the existing wells to draw cross sections of the subsurface. These cross sections show the depths and thicknesses of source rocks and reservoir rocks.

Geophysicists use three methods of oil exploration: magnetic, gravity, and seismic exploration. Magnetometers are used to measure Earth's magnetic field at various locations, while gravity meters measure gravity. Variations in either the magnetic field or gravitational force can indicate petroleum traps hidden below.

During seismic exploration, sound is transmitted into the subsurface using dynamite or a vibrator truck that shakes the ground. As the sound passes through the subsurface, sound waves reflect off layers of sedimentary rocks. The reflections are recorded by detectors called geophones or jugs, which are spread out on the surface in a geometrical pattern. Computers are used to process the seismic data to create a picture of the subsurface rock layers.

Seismic exploration

© 2012 Encyclopædia Britannica, Inc.

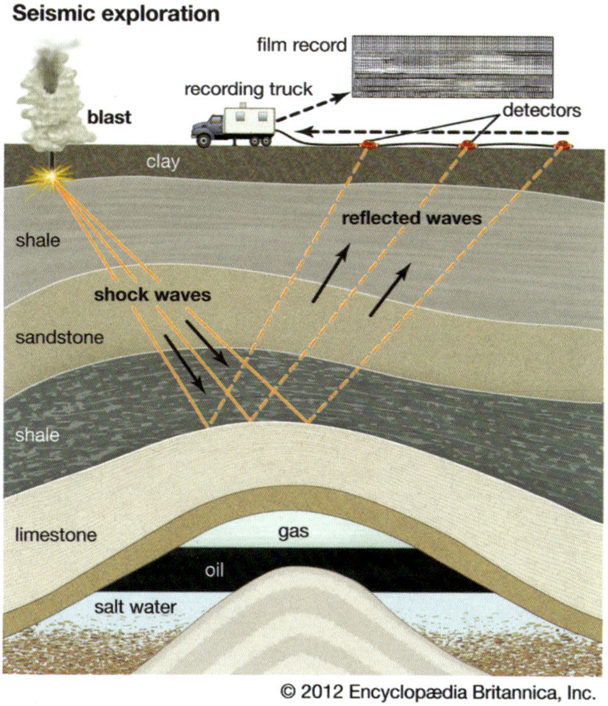

In seismic exploration, sound impulses are put into the ground and the echoes are recorded to image the subsurface rock layers and find petroleum traps.

OIL WELLS

Because drilling rigs are very expensive to operate, they are kept drilling twenty-four hours a day. On

Offshore, all the production equipment is located on the deck of a production platform, which is either fixed or floating.

land, there are three eight-hour shifts each day. Offshore, there are two twelve-hour shifts each day. The driller, usually the most experienced crewmember, operates the machinery and gives orders. Three or more workers called roughnecks do general labor on the floor of the drilling rig.

Each drilling rig is rated for the maximum depth it can drill. The deeper the well, the stronger the drilling rig has to be to support the drillstring in the well. Shallow wells are drilled with small drilling rigs and are relatively inexpensive. Deep wells (below 10,000 feet [3,000 m]) are drilled with large drilling rigs and are very expensive. Offshore exploratory wells are drilled from platforms with legs that stand on the ocean floor in shallow water (jack-up rigs) or float in deeper water (semisubmersibles and drillships).

PRODUCTION AND TRANSPORTATION

The most efficient and economical method to transport oil across the ocean is by crude tankers. A supertanker can typically hold 2 million barrels (320,000 cubic meters) of oil and can be up to 1,500 feet (450 m) long. Modern tankers are built with double hulls separated by a space. If the tanker runs aground and the outer hull is ruptured, the crude oil is contained in the inner hull.

The most efficient way to transport crude oil on land is through pipelines. A gathering system of interconnected pipes conducts oil from wells to a larger-diameter trunk pipeline. The trunk pipeline takes oil to a refinery, port, or storage area. The pipeline system is controlled from one central control room. In that room the pipeline and the flow through it are monitored, and valves anywhere along the system can be remotely opened or closed.

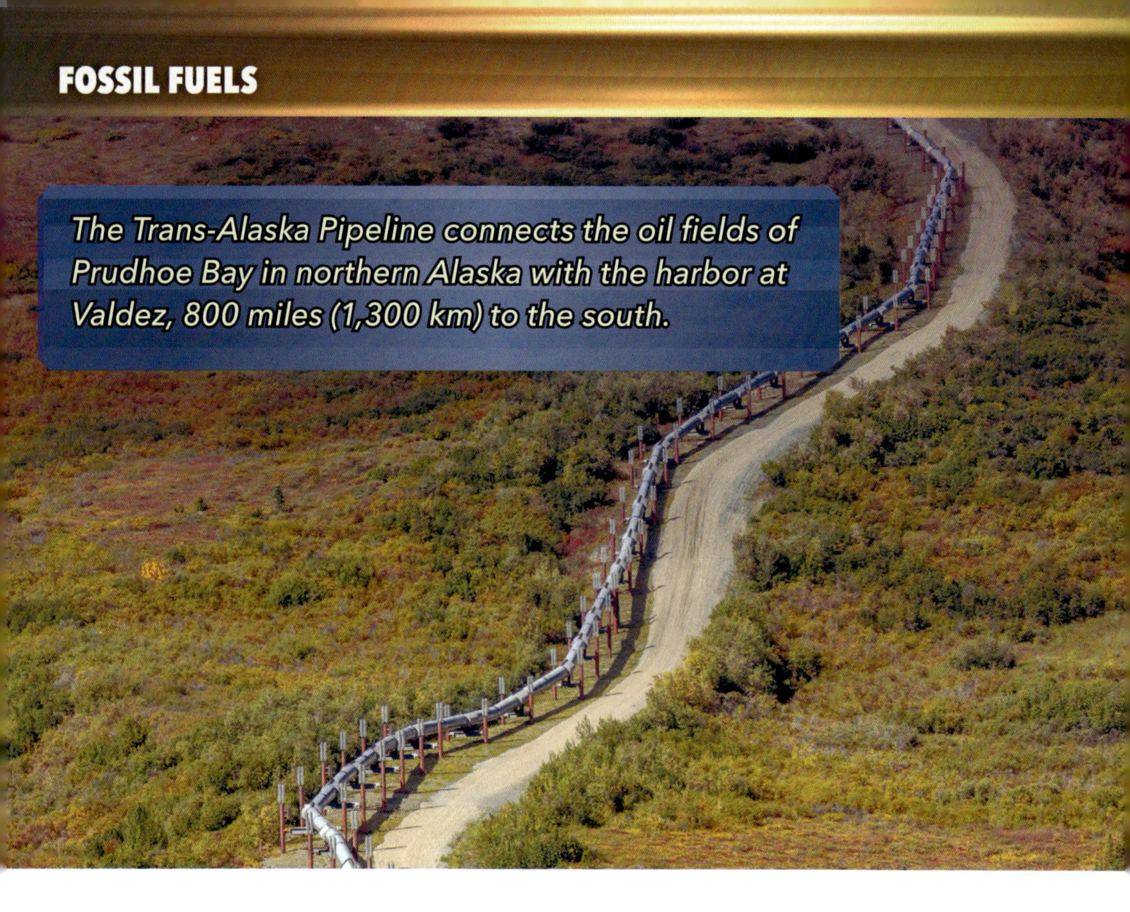

The Trans-Alaska Pipeline connects the oil fields of Prudhoe Bay in northern Alaska with the harbor at Valdez, 800 miles (1,300 km) to the south.

Many smaller oil fields are not connected to a refinery by pipeline. When the stock tanks are filled, the crude oil is transferred to a tanker truck, which takes the oil to a refinery.

In a refinery, crude oil is separated into useful products such as gasoline, jet fuel, diesel fuel, home heating fuel, lubricating oils, and asphalt. Each refinery is uniquely configured to process a specific raw material into a desired slate of products.

CHAPTER THREE

NATURAL GAS

Natural gas is a mixture of flammable gases, mainly the hydrocarbons methane and ethane, that occurs beneath Earth's surface. Helium is also found in relatively high concentrations in natural gas. Natural gas usually occurs in association with petroleum because geological conditions favorable for it generally are favorable for natural-gas occurrence as well. Although many natural gases can be used directly from the well without treatment, some must be processed to remove such undesirable elements as carbon dioxide, hydrogen sulfide, and other sulfur components.

Residential and commercial uses consume the largest proportion of natural gas in both North America and Western Europe, while industry consumes the

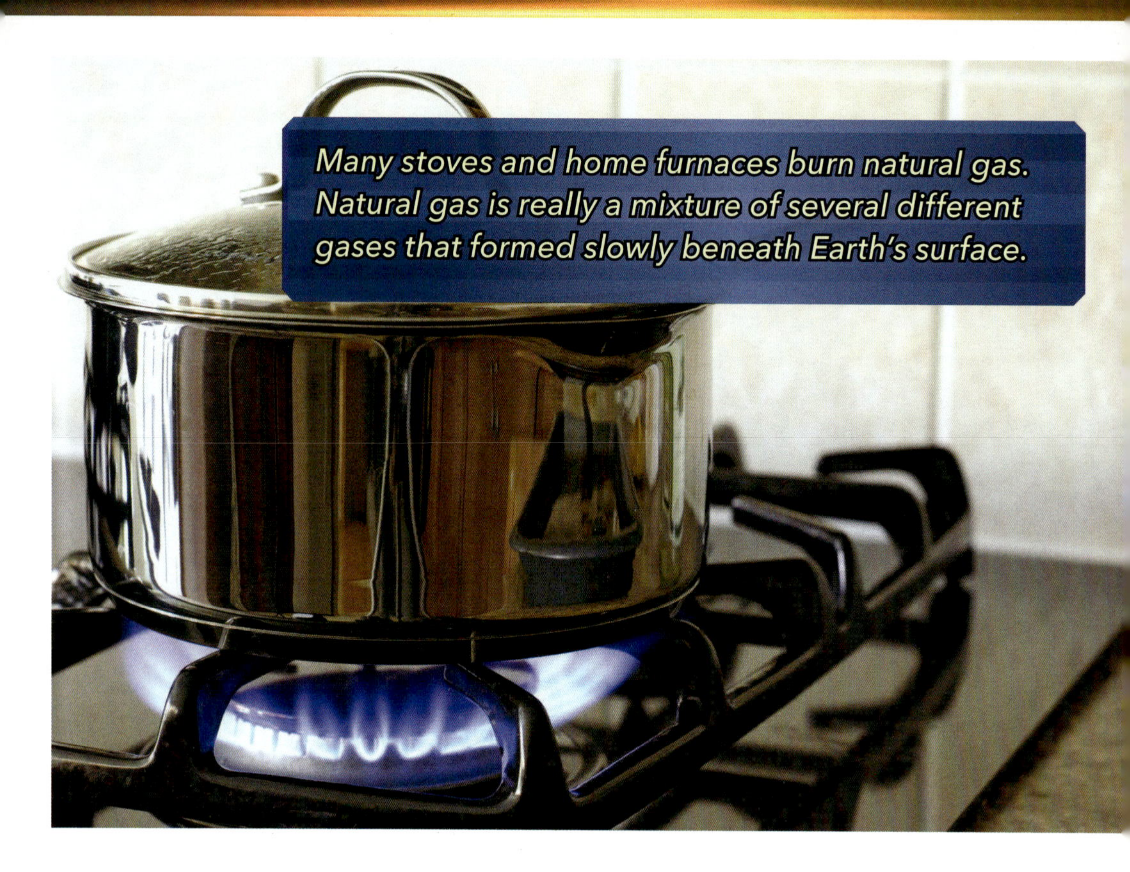

Many stoves and home furnaces burn natural gas. Natural gas is really a mixture of several different gases that formed slowly beneath Earth's surface.

next largest amount and electric-power generation is a distant third in natural-gas consumption. By far the major use of natural gas is as fuel, though increasing amounts are used by the chemical industry for raw material.

HOW NATURAL GAS FORMS

Freshly deposited sediments at the bottom of seas and lakes are the site of intensive bacterial activity that is capable of producing methane, carbon dioxide, nitrogen, and nitrogen oxide (NO) from the organic matter in the sediment. Although methane is formed in abundance (as, for example, marsh gas and peat gas), ethane and heavier hydrocarbons are almost entirely missing. With burial of the sediment beneath succeeding deposits, bacterial activity ceases, and the organic matter is transformed into kerogen, an insoluble product with a complex macromolecular structure.

Most petroleum and natural gas are produced by the degradation of kerogen within the Earth by naturally occurring heat. The sediments containing the organic matter that produces gas are clays or fine limestones, which are mainly compact and relatively impermeable. The gas is expelled and migrates to reservoir levels, which are made up primarily of sands and sandstones and are porous and permeable.

Gas inside the reservoir circulates in the porous space—for example, between sand grains. Normally the pores are filled by water, but gas, because of its very much lower specific gravity, tends to occupy the upper parts of the reservoir level, whereas water remains in the lower parts. For gas to accumulate, the gas must be trapped; that is, the reservoir must be sealed at the top by an impermeable stratum or cap rock, such as clay or salt, with the entire reservoir-cover

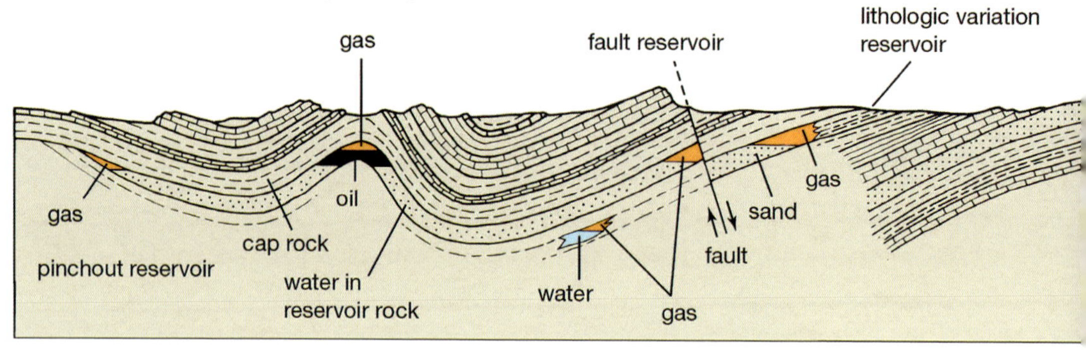

© 2011 Encyclopædia Britannica, In

For gas to accumulate in an underground reservoir, it must be trapped. This means the reservoir must be sealed at the top by an impermeable cap rock of clay or salt.

structure shaped in such a way as to prevent gas from leaking to the surface.

Among the largest accumulations of natural gas are those of Urengoy in Siberia, the Texas Panhandle in the United States, Slochteren-Groningen area in the Netherlands, and Hassi R'Mel in Algeria. Gas accumulations are mostly encountered in the deeper parts of sedimentary basins. On the Gulf coast of the United States more than half of the accumulations of petroleum and natural gas discovered at depths greater than 11,800 feet (3,600 m) are gas fields.

PROCESSING AND TRANSPORTATION

Natural gas is not directly usable when it comes from the well. Often it is extremely rich in methane and contains heavy hydrocarbons. In addition, it contains varying proportions of nitrogen and carbon dioxide, hydrogen sulfide, and other sulfur compounds. The aim of gas processing is to make the gas suitable for various uses and to recover the liquid or solid components, which are often of greater commercial value than the gas itself.

After its arrival at the central plant, the gas must undergo the following series of treatments: (1) drying and removal of condensates; (2) sweetening, which removes the hydrogen sulfide and carbon dioxide; (3) removal of other sulfur compounds; (4) stripping, which extracts the hydrocarbons with longer chains than methane and produces a gas conforming to commercial specifications. These operations are performed continuously and require an extremely sophisticated control system.

Gas fields are often located far from the major centers of consumption. Consequently, the gas must be transported, though refineries are frequently installed near the vicinity of production fields. Transportation of

This Patterson-UTI drilling rig operated by Ultra Petroleum Resources is extracting natural gas from the Pinedale Anticline natural gas field, near Pinedale, Wyoming.

natural gas depends upon its form. In a gaseous form it is transported by pipeline under high pressure, and in a liquid form it is transported by tanker ship.

Large gas pipelines enable gas to be transported over great distances. Examples are the North American pipelines, which extend from Texas and Louisiana to the Northeast coast, and from the Alberta fields to the Atlantic seaboard.

Transportation pressure is generally 1,000 pounds per square inch (70 kilograms per square centimeter) because transportation costs are lowest for pressures in this range. Pipeline diameters for such long-distance transportation increased from an average of about 24 to 29 inches (60 to 70 centimeters) in 1960 to about

This natural gas pipeline is being upgraded. Almost all overland transportation of natural gas is by pipeline. Transporting natural gas by other modes is more dangerous and expensive.

STORAGE

Land storage of low-temperature liquefied gas requires double-walled tanks with special insulation. Such tanks may hold as much as 1,766,000 cubic feet (50,000 cubic meters). Even larger storage facilities have been created by using depleted oil or gas reservoirs near consumption centers or by the creation of artificial gas fields in aquifer layers. The latter technique developed rapidly, and the number of storage facilities of this type in the United States increased tremendously after 1950. There are also such underground storage areas in France and Germany.

4 feet (1.2 m). Some projects involve diameters of more than 6 1/2 feet (2 m). Because of pressure losses, the pressure is boosted every 50 or 60 miles (80 or 100 km) to keep a constant rate of flow.

Petroleum prospecting has revealed the presence of large gas fields in Africa, the Middle East, Alaska, and elsewhere. Gas is transported from such areas by ship. The gas is liquefied to –256°F (–160°C) and transported in specially designed tankers.

Chapter Four

THE DOWNSIDES OF FOSSIL FUELS

Today fossil fuels supply more than 80 percent of all the energy consumed by the industrially developed countries of the world. While they underpin the world economy, fossil fuels are not without their downsides. The main disadvantages of fossil fuels are their limited supply on Earth and the environmental harm they cause.

LIMITED SUPPLY

Fossil fuels are nonrenewable. Once exhausted, they can never be replaced. For years coal was mined as though it were inexhaustible—about one ton wasted for each ton mined. Various government agencies have promoted more efficient mining methods. In addition, the use of other sources of energy in home

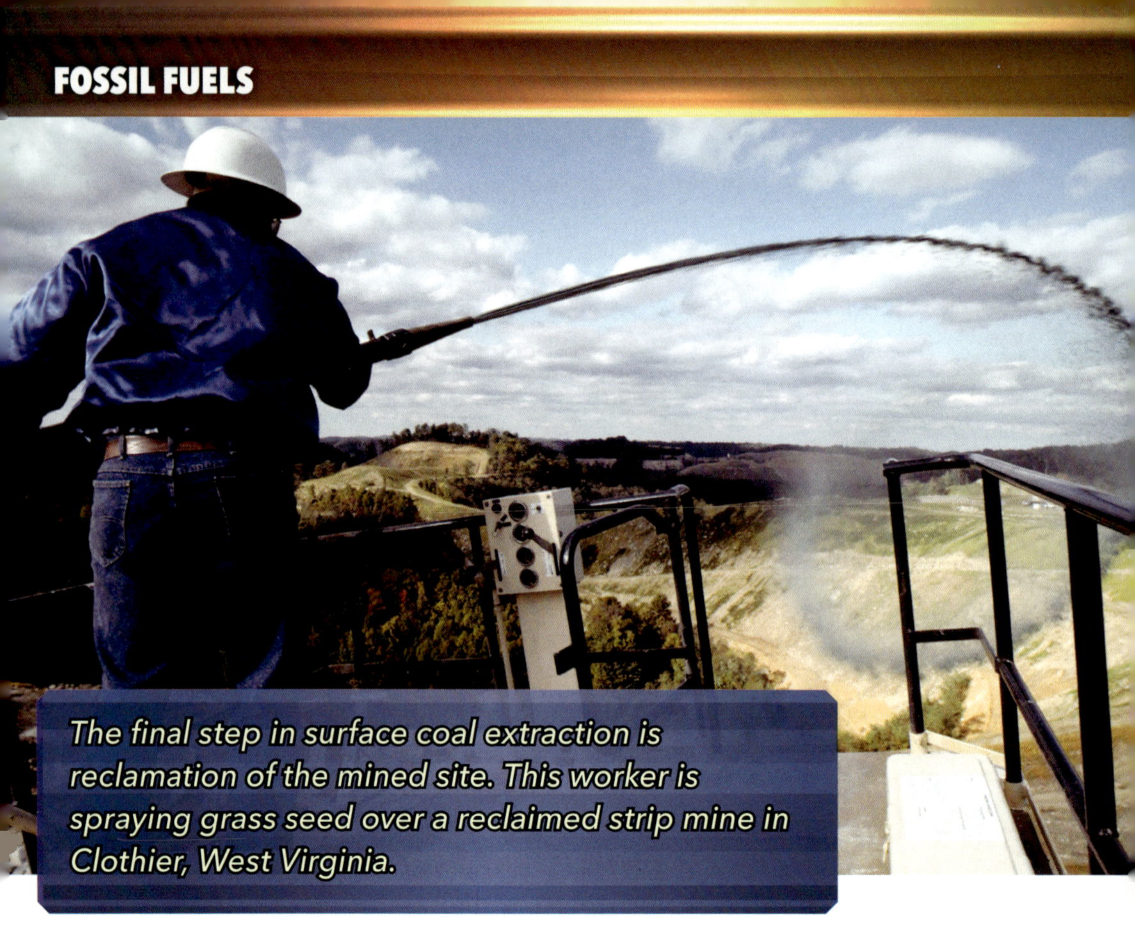

The final step in surface coal extraction is reclamation of the mined site. This worker is spraying grass seed over a reclaimed strip mine in Clothier, West Virginia.

construction and in industry, and to generate electricity, has greatly extended the life of the coal supply. Natural gas and petroleum were once carelessly wasted also. In earlier days, for example, because no use for natural gas had been found, it was burned off or allowed to escape into the air.

ENVIRONMENTAL PROBLEMS

Without proper land reclamation programs, surface mining for coal can strip off the fertile topsoil in an area and leave behind an abandoned wasteland in which few, if any, plants can grow. Coal burning can create pollution problems, not only in the atmosphere but also for organisms, especially amphibians and fish, living in aquatic systems.

The leakage of petroleum onto the surface of a large body of water is an oil spill. Oil spills are chiefly the result of intensified petroleum exploration on the continental shelf and the use of supertankers. The total annual release of oil spills exceeds one million tons (907,000 metric tons). The negligent release

DEEPWATER HORIZON

The largest marine oil spill in history was the Deepwater Horizon oil spill. The spill began after an explosion occurred on April 20, 2010, on the Deepwater Horizon oil rig in the Gulf of Mexico. Over the next five months, almost 5 million barrels of oil were discharged into the gulf waters before the leak was sealed. (Five million barrels of crude oil equal approximately 682,000 metric tons.) That oil spill damaged an estimated 1,100 miles (1,770 km) of shoreline, affecting thousands of birds, mammals, and sea turtles.

This cleanup crew pressure-washed rocks coated with oil from the Exxon Valdez *oil spill, in Prince William Sound, Alaska. The* Exxon Valdez *spill occurred on March 24, 1989.*

of used gasoline solvents and crankcase lubricants by industries and individuals aggravates the problem. The costs of oil spills are considerable in both economic and ecological terms. Oil spills are harmful to birds and many forms of aquatic life, and no thoroughly satisfactory cleanup method has yet been developed.

Fracking is a technology used to access natural gas or crude oil that is trapped underground in rock called shale. To remove the gas or oil, a hole is drilled through many layers of rock until the shale is reached. A fluid made up of water, sand, and chemicals is injected into the shale. This opens up cracks in the rock, which allows the trapped gas or oil to flow through a pipe to the surface. Fracking has raised many environmental concerns. The wastewater that results from fracking is highly polluted. It is full of chemicals and harmful elements. When this waste is disposed of incorrectly, it enters waterways. Another environmental concern is the earthquakes that have been detected in connection with fracking.

GLOBAL WARMING

Burning fossil fuels releases the greenhouse gas carbon dioxide into the air. Greenhouse gases absorb and trap heat emitted from Earth's surface through a process known as the greenhouse effect. About three-fourths of human-caused greenhouse gases come from the burning of fossil fuels. Automobiles, factories, and coal-burning power plants are the biggest producers of this gas.

The dramatic increase in greenhouse gases since the Industrial Revolution and their effect on global climate are of great concern to scientists and others. A report released by the Intergovernmental Panel on Climate Change in 2014 projected a 4.7–8.6°F

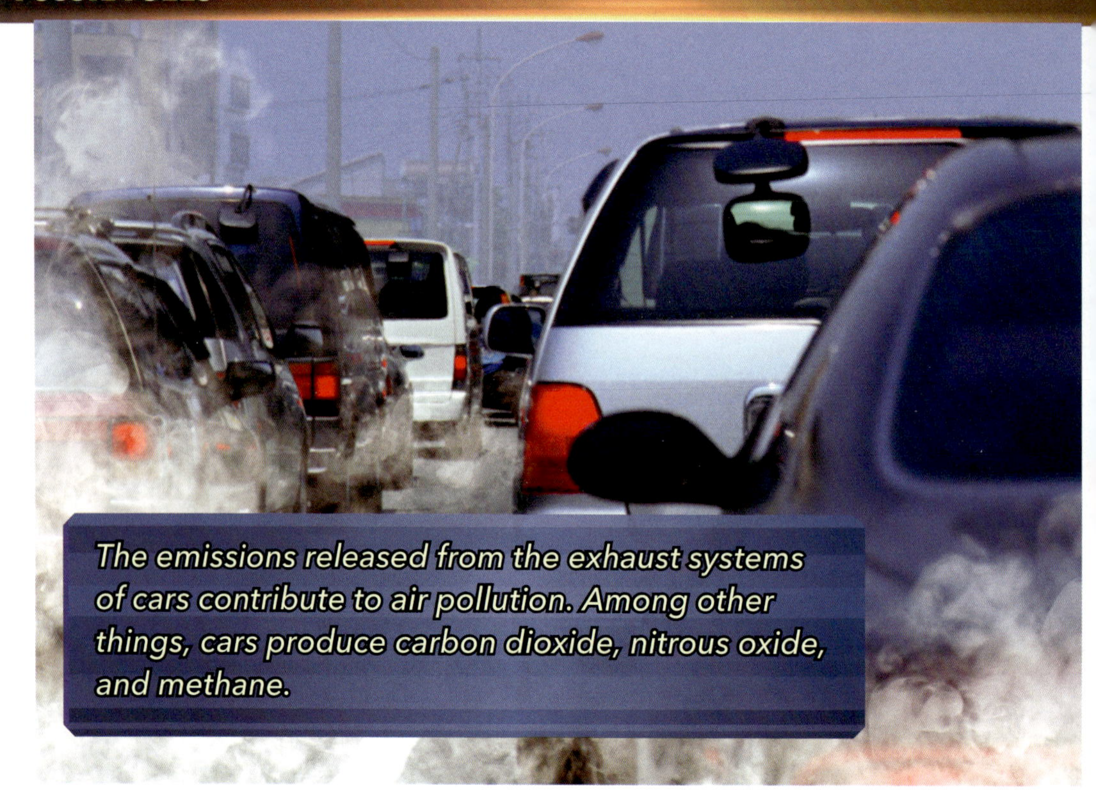

The emissions released from the exhaust systems of cars contribute to air pollution. Among other things, cars produce carbon dioxide, nitrous oxide, and methane.

(2.6–4.8°C) rise in average surface temperature by 2100 if humans do not make major changes to reduce greenhouse gas emissions.

Should present trends in the emissions of greenhouse gases, particularly carbon dioxide, continue, climatic changes larger than any experienced by modern human civilization are expected. This could substantially alter natural and agricultural ecosystems, human and animal health, and climate patterns, increasing the incidence of droughts and major storms and disrupting living things and their interrelationships.

The San Gorgonio Pass Wind Farm in California is the oldest wind farm in the United States. Wind power is one of several alternative energy sources that could replace fossil fuels.

Worldwide, growing concern about global warming in the early twenty-first century led to greater efforts to reduce the use of fossil fuels and to rely increasingly on renewable, nonpolluting energy sources such as solar, wind, and geothermal power. New methods of using biomass energy, by burning plant matter or producing ethanol and other fuels from crops and agricultural wastes, were also investigated, and more electric and gasoline-electric hybrid automobiles were introduced.

GLOSSARY

ANTICLINE An arch of rock arranged in layers that bend downward in opposite directions from the top.

ATOM One of the tiny particles that are the basic building blocks of all matter. Atoms can be combined with other atoms to form molecules, but they cannot be divided into smaller parts by ordinary means.

COLLOIDAL Consisting of a very finely divided substance which is scattered throughout another substance.

COMBUSTION A chemical reaction that produces heat and light, such as fire.

CONTINENTAL SHELF An area, of varying distance from shore, that lies in water of relatively shallow depth.

DRILLSHIP A ship equipped for drilling (as for oil) in the ocean floor.

GREENHOUSE EFFECT The warming of Earth's surface and the air above it, caused by gases in the air that trap energy from the sun.

HYDROCARBON An organic compound containing only carbon and hydrogen.

INSOLUBLE Impossible or difficult to dissolve.

JACK-UP RIG A mobile offshore drilling unit that rests on the ocean floor.

KEROGEN A complex waxy mixture of hydrocarbon compounds that is the primary organic component of oil shale.

ORGANIC Of, relating to, or containing carbon compounds.

PEAT A dark brown or blackish material that is the remains of plants partly decayed in water and is sometimes dug up and dried for use as fuel.

PERMEABLE Having pores or openings that permit liquids or gases to pass through.

REFINERY A place where oil is refined, so that it is in an easily usable from.

SEDIMENT Material (as stones and sand) deposited by water, wind, or glaciers.

SEISMIC Relating to an earthquake or other earth vibration (such as an explosion).

SEMISUBMERSIBLE An offshore drilling platform with a hull that is entirely underwater, while the operational platform is held well above the surface on supports.

STRATUM (plural **STRATA**) One of a series of layers.

SYNTHETIC Produced artificially.

TRAP A large, bowl-shaped space in reservoir rock that traps huge amounts of petroleum near Earth's surface.

TURBINE An engine whose central driving shaft is fitted with a series of blades spun around by the pressure of a fluid (as water, steam, or air).

VISCOUS Having the characteristic of stickiness.

FOR MORE INFORMATION

Belton, Blair. *How Coal Is Formed* (From the Earth: How Resources Are Made). New York, NY: Gareth Stevens Publishing, 2017.

Chambers, Catherine. *How Harmful Are Fossil Fuels?* (Earth Debates). Chicago, IL: Heinemann Raintree, 2015.

Cunningham, Anne C. *Critical Perspectives on Fossil Fuels vs. Renewable Energy* (Analyzing the Issues). New York, NY: Enslow Publishing, 2017.

Dickmann, Nancy. *Burning Out: Energy from Fossil Fuels* (Next Generation Energy). New York, NY: Crabtree Publishing Company, 2016.

Doeden, Matt. *Finding Out about Coal, Oil, and Natural Gas* (Searchlight Books: What Are Energy Sources?). Minneapolis, MN: Lerner Publications Company, 2015.

Goldstein, Margaret J. *Fuel under Fire: Petroleum and its Perils*. Minneapolis, MN: Twenty-First Century Books, 2016.

Gordon, Sherri Mabry. *Out of Gas: Using Up Fossil Fuels*. New York, NY: Enslow Publishing, 2016.

Idzikowski, Lisa. *Pipelines and Politics* (At Issue: American Politics). New York, NY: Greenhaven Publishing, 2018.

Iyer, Rani. *Endangered Energy: Investigating the Scarcity of Fossil Fuels* (Fact Finders: Endangered Earth.) North Mankato, MN: Capstone Press, 2015.

Labrecque, Ellen. *Drilling and Fracking*. Ann Arbor, MI: Cherry Lake Publishing, 2017.

Machajewski, Sarah. *20 Fun Facts About Earth's Resources* (Fun Fact File: Earth Science). New York, NY: Gareth Stevens Publishing, 2018.

Macken, JoAnn Early. *Take a Closer Look at Oil*. South Egremont, MA: Red Chair Press, 2016.

Mara, Wil. *Inside the Oil Industry* (Big Business). Minneapolis, MN: Essential Library, 2017.

Nagelhout, Ryan. *How Natural Gas Is Formed* (From the Earth: How Resources Are Made). New York, NY: Gareth Stevens Publishing, 2017.

Nelson, Kristen Rajczak. *How Oil Is Formed* (From the Earth: How Resources Are Made). New York, NY: Gareth Stevens Publishing, 2017.

Streissguth, Tom. *Inside the Coal Industry* (Big Business). Minneapolis, MN: Essential Library, 2017.

Wang, Andrea. *How Can We Reduce Fossil Fuel Pollution?* Minneapolis, MN: Lerner Publications, 2016.

WEBSITES

Environmental and Energy Study Institute
http://www.eesi.org/topics/fossil-fuels/description
Facebook, Twitter, YouTube: @eesionline

International Energy Agency
https://www.iea.org/tcp/fossilfuels/
Facebook: @internationalenergyagency;
Twitter: @IEA

U.S. Department of Energy
https://energy.gov/science-innovation/energy-sources/fossil
Facebook: @energygov;
Instagram, Twitter: @energy

U.S. Environmental Protection Agency
https://www.epa.gov/nutrientpollution/sources-and-solutions-fossil-fuels
Facebook, Twitter: @EPA

INDEX